COMMON SNAPPING TURTLE AS PET

Complete owners guide to common snapping turtle, care, reproduction, management and many more included

ALBERT A. NELSON

Table of Contents

INTRODUCTION...5

CHAPTER 1 7

INTERESTING FACTS AND COMMON CHARACTERISTICS OF TURTLE

HATCHLINGS...7

CHAPTER 2 14

CREATING AN ENVIRONMENT SUITABLE FOR COMMON SNAPPING

TURTLES 14

CHAPTER 3 24

FEEDING GUIDELINES AND COMMON SNAPPING TURTLE DIETS..24

CHAPTER 4 35

SNAPPING TURTLE HEALTH ISSUES: UNDERSTANDING CARE AND

RISKS 35

CHAPTER 5 49

COMMON TURTLE HANDLING DOS AND DON'TS......................49

CHAPTER 6 ....**61**

BREEDING COMMON SNAPPING TURTLES: A PET OWNER'S GUIDE **61**

CHAPTER 7 ....**74**

COMMON LOGGERHEAD TURTLE SPECIES SPOTLIGHT: UNIQUE CHARACTERISTICS **74**

CHAPTER 8 ....**86**

AQUATIC ADVENTURES: SETTING UP AN ENRICHMENT TANK FOR YOUR TURTLE **86**

CHAPTER 9 ....**98**

MYTHS ABOUT COMMON SNAPPING TURTLES: SEPARATING FACT FROM FICTION **98**

CHAPTER 10 ....**108**

EDUCATIONAL ACTIVITIES FOR CHILDREN: COMMON SNAPPING TURTLES **108**

QUESTIONS TO BE ASKED **124**

Introduction

Welcome to the wonderful world of the common snapping turtle, where classic charm meets aquatic wonders. Known for their unusual appearance and ferocious behavior, these fascinating reptiles have captured the hearts of both reptile lovers and pet owners. In this exploration we embark on a quest to discover the secrets of these ancient creatures by diving into their habits, environment and the art of proper care.

The common snapping turtle, Chelydra serpentina, has a hard exterior with a black, scaly shell and a strong pair of dignified teeth. As we navigate the complexities of their physiology, behavior and unique characteristics, you will develop a comprehensive understanding of what makes these turtles truly unique.

From building a comfortable environment for your common snapping turtle to unraveling the mysteries of their appetites, we'll walk you through the intricacies of ethical pet keeping. Learn the dos and don'ts of caring for these turtles to ensure your safety as well as the health of your furry friend.

Discover the intricacies of creating new life among these ancient reptiles by entering the breeding field. This complete guide is your key to understanding the mysteries of the Common Snapping Turtle, including insights into their health care needs, species diversity, and debunking common myths.

So come along on this educational journey as we explore the aquatic adventures of caring for, understanding and appreciating the amazing common snapping turtle.

Chapter 1

Interesting facts and common characteristics of turtle hatchlings

The common snapping turtle (Chelydra serpentina) is an example of prehistoric lineage survival, attracting admirers for its striking beauty and curious behavior. In this extensive investigation, we peel back the layers of interest surrounding these ancient reptiles, delving into a treasure trove of data and shedding light on their diverse traits and behaviors.

The Wonders of Evolution: A Retrospective

The common snapping turtle, also known as a living fossil, has undergone minor evolutionary modifications over the ages. These reptiles evolved over 40 million years ago and developed survival techniques, adapting to the diverse conditions in North America. Their unique features, including carapace and strong jaw structure, highlight their unusual evolutionary path.

Anatomy and adaptations are two examples of physical wonders.

Common snapping turtles are easily recognized by their black, muscular shells and long, serpentine necks. Their strong legs and webbed feet give them dexterity in water, allowing them to move through water and land with ease. But their defining feature is their huge jaws, which can deliver a powerful bite and act as a formidable defense mechanism against potential attackers.

Living in harmony in the habitat: where common snapping turtles survive

Common snapping turtles have an adaptable lifestyle, living in a variety of aquatic habitats. These turtles are adept at adapting to conditions, from slow-moving streams to stagnant ponds. Understanding their native habitat is important for pet owners, as it helps create a suitable environment in captivity. Examining complex water quality, habitats and nests reveals the secrets to creating a successful ecosystem for these reptiles.

Feeding preferences: Carnivorous palette

The complexities of the common snapping turtle's diet reflect its carnivorous nature. Fish, amphibians, insects and small mammals are present in the diet, showing a flexible feeding strategy. As pets, supplementing this broad diet will be critical to their health. This class will delve into the art of providing a balanced and healthy diet while ensuring that these unique nutritional needs of turtles are met in captivity.

Reproductive Behavior Revealed: The Mysterious Mating Rituals

Explore the mating rituals of the common Snapping Turtle, where courtship is observed and aberrant behaviors are laid down for future generations. This section takes an in-depth look at the current complexities of breeding, the importance of size and age at mating, and the precautions required for effective captive breeding. Understanding the complexity of this process is important for those interested in participating in responsible reproductive practices.

Health Care Essentials: Shell Shock

A complete guide to health care essentials reveals the keys to keeping your common snapping turtle healthy. This class teaches pet owners how to be proactive caregivers, from recognizing the signs of illness to providing proper medical care. Topics covered include shell health, common diseases and prevention strategies to help these amazing reptiles live long and healthy lives.

Aquatic enrichment for captive turtles: Exploring the water

Creating a stimulating and enriching environment for common snapping turtles in captivity requires careful planning. Delve into the art of creating a well-designed tank that reflects their natural habitat. Enrichment activities are important for cerebral stimulation, and this class provides imaginative ways to engage your turtle both physically and intellectually.

Careful Handling: Do's and Don'ts

Due to their protective nature, common snapping turtles require a careful approach when handling. This class teaches safe handling procedures while respecting the turtle's instincts. From reading body language to giving a strong hug, learn what to do and what to do with these delightful creatures.

Common snapping turtle species in the spotlight

Dive into the world of diversity of the common Snapping Turtle by learning about the different breeds and species. Discover the subtle differences in size, appearance and habitat preferences that distinguish these reptiles. A better understanding of the species' focus will broaden your appreciation of the rich tapestry of biodiversity in common snapping turtles.

Myth busting: separating fact from fiction

In order to promote accurate information and appropriate pet ownership, it is important to dispel myths about snapping turtles. This episode serves as a myth-busting guide, addressing misconceptions about obesity and dispelling nutrition myths, arming enthusiasts with factual knowledge.

Educational Adventures for Kids: Common Snapping Turtles

With educational activities focused on common snapping turtles, you can inspire a new generation of environmentalists. This class offers fascinating, age-appropriate lessons that promote understanding of these reptiles and their role in the ecosystem. These educational trips make learning about turtles a fun and memorable experience, with interactive games and hands-on activities.

Finally, "Common Snapping Turtles: Fascinating Facts and Unique Features" is a comprehensive reference that invites readers to immerse themselves in these ancient reptiles. Whether you're a veteran or a curious newcomer, this exploration reveals the mysteries and wonders that make common snapping turtles a truly fascinating subject of study and admiration.

Chapter 2

Creating an environment suitable for common snapping turtles

Creating the perfect home for your Common Snapping Turtle (Chelydra Serpentina) is a rewarding endeavor that combines the art of true design with the science of animal care. In this detailed tutorial, we explore the complexities of creating an environment that mimics the native habitats of these ancient reptiles to ensure their physical and mental well-being.

Understanding Natural Habitats: A Guide to Design
In the wild, common small turtles live in a variety of habitats, from slow-moving streams and ponds to marshes and swamps. In order to maintain a stress-free and prosperous existence, it is essential that they replicate the beneficial properties of their environment

in captivity. Start by researching your turtle species or subspecies to personalize the habitat for its needs.

Water Bodies: Create a large body of water using clean, non-chlorinated water. Snapping turtles are excellent swimmers, and a large swimming area allows them to exercise and explore. Include a mixture of sand and gravel to create a surface that can be drilled and drilled like a natural substrate.

Perching Zones: Provide your turtle with multiple feeding areas, such as large flat rocks or floating platforms, that extend from the water to the thermowell. Adequate roosting areas are critical to your turtle's overall health and digestion.

Hiding and terrain units: Include terrain units with hiding places to simulate the varied topography of their natural environment. Create a varied and attractive habitat with driftwood, rocks and plants. Make sure parts of the surface are easily accessible so that your turtle can move between water and land.

Water Quality: Invest in a reliable filtration system to maintain good water quality. Turtle care requires regular water changes, water testing, and temperature control. Clean, well-filtered water is essential to the health and longevity of the Common Snapping Turtle.

Environmental enrichment: Mind and body stimulation
Creating a stimulating habitat requires engaging your turtle's cognitive and behavioral components in addition to meeting basic physical needs. Environmental enrichment is critical to encouraging natural activities, reducing stress, and reducing boredom.

Aquatic Plants: Incorporate live or artificial aquatic plants to enhance the beauty of the ecosystem. Live plants improve water quality by absorbing nitrates and providing protection for your turtle. Make sure any artificial plants your turtle comes in contact with are secure.

Floating objects: Include floating objects such as cork bark or large, floating leaves. These act as nesting platforms, but also allow your turtle to explore and interact with its environment.

Interactive Feeders: Use puzzle feeders or sprinkle food on the habitat to encourage your turtle to eat and exercise its natural hunting instincts. This not only stimulates the mind, but also adds something interesting to eat.

Variety of textures: Create areas with different textures, such as sand, gravel, or a combination of the two. This allows your turtle to engage in natural digging habits, providing both mental and physical stimulation.

Creating good air temperature and light conditions

Maintaining proper temperature and lighting conditions is critical to the health of your common snapping turtle. These reptiles are ectothermic, which means they depend on external heat to regulate their body temperature.

Seat temperature: Make sure the heated area reaches the proper temperature, which is normally between 85-90°F (29-32°C). To provide a warm place for your turtle to regulate its temperature, use an incandescent lamp.

Install a UVB light source to replicate natural sunlight that promotes healthy calcium metabolism and prevents metabolic bone disease. UVB exposure is important for your turtle's overall health and shell development.

Create thermal efficiency in the dwelling, away from the cooking area in cooler areas. This allows your turtle to choose its preferred temperature, moving between hot and cold zones as needed.

Regular cleaning and hygiene care

A clean and well-maintained habitat is critical to the health of your common snapping turtle. Regular cleaning practice reduces the accumulation of harmful microorganisms and ensures a sanitary environment.

Filter System: Invest in a reliable filter system suitable for your turtle's habitat. In order for the water to be clean and transparent, mechanical and biological filter units work together.

Water Changes: Do regular partial water changes to remove debris and maintain water quality. The frequency of watering depends on the size of the habitat and the number of turtles.

Site cleaning involves removing uneaten food, waste, and debris from water and land environments. Regular monitoring helps prevent the accumulation of dangerous substances.

Health Monitoring: Watch for signs of stress
It is important to observe your common snapping turtle daily for any signs of discomfort or health problems. Note any changes in behavior, appetite, or physical appearance. Regular health checkups with a veterinarian are recommended for early detection of potential problems.

Behavioral Analysis: Note any changes in swimming style, swimming habits or activity level. Abnormal behavior can be a sign of stress, illness or disease.

Track your turtle's eating habits with an appetite monitor. A sudden loss of appetite, not eating, or irregular stools can be signs of underlying health problems.

Shell and Skin Inspection: Check the shell and skin regularly. Any abnormalities such as shell deformations, discoloration, or lesions should be immediately checked by a veterinarian.

Legal considerations and ethical practices in responsible pet ownership

Before getting a Common Snapping Turtle as a pet, learn about the ownership laws in your area. Some authorities may impose special laws or restrictions to protect both the turtles and the ecosystem. Also, make sure the turtle was not caught illegally from the wild and choose reputable breeders or rescue organizations.

Protection and awareness in education

In addition to the personal satisfaction of caring for a common snapping turtle, consider the broader impact of your actions by promoting conservation and awareness. Participate in education by sharing your knowledge and experiences. Raise awareness of the importance of conserving these beautiful reptiles and their habitats, thus contributing to biodiversity conservation.

Ultimately, designing a suitable environment for your common snapping turtle is a multifaceted task that combines scientific concepts and a deep appreciation for the natural world. By incorporating knowledge of their native habitats, enriching the environment, maintaining favorable conditions, and practicing ethical pet care, you can give your turtle a home that mimics the beauty and complexity of its natural environment. This dedication will not only improve your turtle's health, but also create a deeper bond between humans and these ancient and amazing creatures.

Chapter 3

Feeding Guidelines and Common Snapping Turtle Diets

Feeding the common snapping turtle (Chelydra serpentina) is about providing a balanced diet while also being aware of the nutritional requirements of these interesting reptiles. In this detailed guide, we'll explore the intricacies of your common snapper's diet, nutritional needs, and nutritional concerns for health and safety issues.

Addressing dietary choices with a meat eater's culinary palette

Common snapping turtles are primarily carnivores, selecting a variety of prey items found in their natural environment. Understanding their dietary preferences is the first step in developing a nutritionally appropriate diet plan.

Protein-rich diet: Common snapping turtles thrive on a protein-rich diet. Fish, amphibians, insects, crabs and even small mammals are on their menu. Replacing this protein gap in pets is critical to their overall health and well-being.

Live Prey vs. From Prepared Foods: A combination of live game and prepared foods provides a balanced diet. A living predator such as feeder fish, insects, and earthworms stimulates the mind and allows natural hunting tendencies to be expressed. The nutritional profile is enhanced by supplementing with commercial turtle pellets supplemented with key nutrients.

Calcium Factors: Calcium is important for the skeletal health of common snapping turtles. Providing calcium-rich foods such as bone fragments, calcium blocks, or game products with a calcium supplement can help prevent metabolic bone disease.

Dietary adjustment for age and size: growth estimation
As common snapping turtles grow and develop, their nutritional needs change. It is important to balance their diet with their age and size to promote healthy growth.

Chicks and Juveniles: Young turtles eat a lot and grow quickly. Small, bite-sized prey items such as small fish, insects and commercial turtle pellets are recommended. At this time, calcium supplementation is essential for a strong shell and bone growth.

Adults: As they grow, common snapping turtles' diets may include larger prey items. Their diet may include whole fish, large invertebrates, and a fair amount of rodents. Maintaining bone health requires balancing protein and calcium intake.

Feeding Frequency and Timing: Finding the Sweet Spot

A consistent feeding schedule encourages healthy eating habits and helps control the turtle's appetite. The frequency and portion size depends on characteristics such as age, size and general health.

Hatchlings and young turtles need to be fed more often, often every day or every other day. This ensures that they get the nutrients they need for growth and development.

Adult Turtles: Adult common snapping turtles should be fed no more than two to three times per week. Party sizes should be adjusted based on the individual turtle's appetite and activity level to avoid overfeeding, which can lead to obesity and other health problems.

Fasting Periods: Implement intermittent fasting periods to approximate a regular diet. This practice helps prevent obesity, and also allows the digestive system to relax.

Variety is the spice of life: dietary variety

A wide and varied diet ensures a wide variety of nutrients and enriches the turtle's life. Experiment with new prey items to keep your mind stimulated and your diet monotonous.

Fish: Offer a variety of fish such as small feeder fish, minnows, or freshwater fish pieces. Eating too much thiaminase-rich fish, such as goldfish, can cause nutritional deficiencies.

Crickets, mealworms, earthworms and other insects provide valuable protein and satisfy the turtle's natural foraging instincts. Make sure that insects given are of acceptable size to avoid choking hazards.

Crustaceans: Regular inclusion of crustaceans such as shrimp or crayfish helps to diversify the diet. These predatory foods are rich in protein and minerals.

Commercial Turtle Pellets: High quality commercial turtle pellets are a practical and nutritionally balanced solution designed exclusively for aquatic turtles. These pellets are often fortified with vitamins and minerals to supplement the turtle's diet.

Hydration is important: increase water consumption
The foods that common snapping turtles eat provide a high level of moisture. But access to safe, chlorinated water is just as important to their health.

Maintain an appropriate amount of water in the habitat for swimming and hydration. The water should be deep enough to get the turtle in, but shallow enough to easily reach the surface.

Spraying and Dipping: Provide opportunities to sprinkle or dunk turtles that cannot easily drink from water food. Allowing your turtle to burrow in a shallow container or gently compacting it will help keep it moist.

Viewing Dietary Indicators: Nutritional Health Indicators It is important to monitor your common snapping turtle's signs of nutritional health to change the diet as needed. Among the main indicators are:

Growth and Development: Assess the growth and development of chicks and juveniles. A balanced diet promotes stable growth, a well-shaped shell and strong legs.

Shell Integrity: The turtle's shell serves as a visible indicator of its overall health. A smooth shaped shell without any abnormalities or discoloration indicates that the shell has received proper nutrition.

Activity Level: A well-fed turtle's activity level is normal. If your turtle seems weak or very inactive, this could indicate malnutrition or other health problems.

Troubleshooting and solutions for joint feeding issues
Feeding problems often occur when caring for snapping turtles. Addressing these conditions as soon as possible will ensure your pet's continued safety.

Reluctance to eat: A brief loss of appetite is common, especially during environmental changes or molting. However, prolonged refusal may indicate underlying health problems. Consult a reptile vet to rule out any underlying issues.

Overfeeding: Overfeeding can lead to obesity, which has a negative impact on the turtle's overall health. Feeding frequency and portion sizes should be adjusted based on the turtle's age, size, and activity level.

Imbalance in Nutrients: Be aware of nutritional imbalances, especially calcium deficiency. Review the turtle's diet regularly and talk to your veterinarian about making any necessary changes or additions.

Consultation with a veterinarian: professional advice
Regular veterinary checkups are critical to the long-term health of your Common Snapping Turtle. A reptile veterinarian can provide customized recommendations, perform health screenings, and address any concerns about your turtle's dietary and nutritional needs.

Regular Vet Checkups: Schedule regular vet checkups to monitor your turtle's overall health. Early diagnosis of potential health problems allows timely intervention.

Nutritional Consultation: For personalized nutritional advice, seek the advice of a reptile vet. They will analyze the nutritional content of your turtle's current diet, provide supplements if needed, and adjust feeding programs to your turtle's individual needs.

Sharing knowledge and spreading awareness through education

Share your knowledge of common snapping turtle nutritional needs with hobbyists and future turtle keepers as a responsible pet owner. Educational outreach can help build a community that respects the welfare of these fascinating reptiles and leads to a better understanding of their care requirements.

Finally, understanding common snapping turtle feeding guidelines and nutritional needs is a flexible and fun part of responsible pet care. By using a holistic approach that takes into account age, size, breed, water level and the importance of expert advice, you can provide your turtle with a nutritionally balanced and rich diet. This not only improves their physical health, but also strengthens their relationship with these delightful old reptiles.

Chapter 4

Snapping Turtle Health Issues: Understanding Care and Risks

Ensuring the health and well-being of your Common Snapping Turtle (Chelydra Serpentina) requires a thorough knowledge of its unique anatomy, common health concerns, and preventive care practices. In this comprehensive tutorial, we'll explore the intricacies of Common Snapping Turtle health, covering everything from routine health checks to dealing with potential complications during captive care.

Key concepts in anatomy and physiology

Before getting into the health care details, it is important to understand the anatomy and physiology of common snapping turtles. These reptiles are known thanks to their hard, scaly carapace, fearsome jaws, and long serpentine necks. Understanding their natural habits, respiratory systems, and thermoregulatory mechanisms lays the foundation for preventive health care.

Common snapping turtles are mostly underwater, although they need to reach the surface to breathe. They have special respiratory organs that allow them to absorb oxygen from air or water. It is very important to provide a well-ventilated environment for their breathing needs.

Snapping turtles rely on external heat sources to maintain their body heat. They bask in the sun to raise their body temperature and then dive into the water to cool off. Maintaining their thermoregulatory balance requires a well-designed housing with adequate cooking areas.

Digestive System: These turtles are carnivores, feeding mainly on fish, amphibians, insects and small animals. Understanding their digestive system helps to create an appropriate diet and identify indicators of digestive system disorders.

Monitor vital signs regularly
Regular health checks allow you to monitor the overall health of your common Snapping Turtle. Physical examinations, behavioral observations, and checking for signs of health problems are part of these evaluations.

Body Condition: Check the turtle's body condition regularly to make sure it is not underweight or overweight. A well-fed turtle is round in shape, but not overly bloated.

Shell health: Look for any abnormalities in the shell, such as distorted shape, color, or shell decay. A healthy bark is smooth and well formed, without damage.

Eyes and Nose: Clear and bright eyes that are free of fluid are signs of good health. Inspect the nose for signs of respiratory problems, such as excessive mucus or foam.

Check the mouth and feet for signs of pain, sores, or unusual swelling. Examine the legs for any injuries, swelling or lameness.

Nutritional considerations: a pillar of preventive care
Proper nutrition is critical to the health and vitality of common snapping turtles. Nutritious, nutritious food promotes their growth, energy levels and immune system.

Provide a varied and nutritious diet that mimics the diversity of their natural prey. Fish, insects, crustaceans, and commercial turtle pellets are all examples. To avoid metabolic bone disease, make sure you get enough calcium.

Overfeeding has a negative effect on the turtle's overall health and should be avoided as it can lead to obesity. Portion sizes should be monitored based on age, size and activity level, and excessive rewards should be avoided.

Hydration: Staying hydrated is important for digestion and overall health. Make sure the turtle has clean, unchlorinated water in which to eat and swim.

Environmental conditions: cleanliness and quality of residence

The health of common snapping turtles depends on keeping their environment clean and well designed. Environmental conditions have a direct impact on their stress levels, immune function and disease susceptibility.

Water Quality: To maintain good water quality, invest in a reliable filtration system. Regular monitoring of water changes and water parameters helps prevent dangerous bacteria and pollution.

Provide plenty of baking and hiding places for temperature control and make sure they are easily accessible. Include hiding places to avoid stress and let the turtle escape when necessary.

Take care of the surface by cleaning and cleaning the debris. Remove uneaten food, feces, and contaminants regularly to avoid the growth of harmful microorganisms.

Avoiding respiratory problems through respiratory health

Common park turtles are prone to respiratory problems, especially when housed in conditions with insufficient or inadequate ventilation. It is important to solve the problem of breathing as soon as possible to avoid problems.

Proper ventilation: Maintain adequate ventilation in the residence to reduce the risk of suffocation. Avoid overfilling the water tank to allow for adequate air circulation.

Watch for signs of respiratory problems, such as wheezing, difficulty breathing through the mouth, or a runny nose. If you notice any of these symptoms, consult a reptile veterinarian for a comprehensive evaluation and proper treatment.

Regular health care and parasite prevention and treatment

Both internal and external parasites can threaten the health of common snapping turtles. Parasite control requires regular monitoring, preventive measures and initial treatment.

Regular fecal examination with a veterinarian is recommended to identify and treat internal parasites. Early detection provides for better intervention.

Quarantine Procedures: Before integrating a new turtle into an established habitat, employ a quarantine phase to look for signs of disease or parasites.

Avoid wild-caught predators: Avoiding wild-caught predators reduces the risk of parasites. Choose raised and commercially raised feedlot animals.

Treatment methods: reduction of anxiety and depression

The common snapping turtle is known for its defensive behavior, and mishandling can cause stress and pain. Adopting gentle and proper handling skills is critical to your turtle's overall well-being.

Limit Handling: Limit unnecessary handling to reduce stress. When handling is necessary, be gentle and confident. Support the turtle's body and avoid using too much force.

Observe defensive behaviors: Snapping turtles are known to snap when confronted. Respect their defensive response and avoid threatening or intimidating them.

Recognizing symptoms in health care

Careful health monitoring requires early detection of signs of disease or illness. Changes in behavior, appetite, appearance, and overall activity level are common signs of health problems.

Track changes in swimming patterns, swimming habits, or overall activity levels. Apathy, excessive hiding, or hostility can all be symptoms of underlying health problems.

Appetite and Weight: A sudden loss of appetite, not eating, or weight loss can indicate a health problem. If you have any concerns about your reptile's feeding habits, consult a reptile veterinarian.

Skin and Shell Abnormalities: Inspect the skin and shell for any unusual abnormalities such as lesions, discoloration, or swelling. Early detection allows timely intervention.

Seeking professional help in medical emergencies

In the event of a medical emergency or serious health problem, it is important to seek professional help quickly. Injuries, respiratory distress and serious illnesses are examples of common emergencies.

Emergency Veterinarian: Find and maintain contact information for a reptile veterinarian who knows how to treat concussed turtles. In an emergency, the quick intervention of professionals can be crucial.

First Aid Knowledge: Learn basic first aid techniques for reptiles. This includes building a temporary hospital enclosure, handling injured turtles carefully, and dealing with frequent emergency situations.

Gender identity and reproductive considerations in reproductive health

Understanding common snapping turtle reproductive health is critical, especially if you have both male and female snapping turtles. Knowing gender differences and providing suitable breeding conditions will help improve their overall well-being.

Gender Identification: Use physical characteristics to distinguish between male and female turtles. Females have shorter tails and smaller claws, while males have longer tails, larger claws, and smaller pointed plates.

Breeding considerations: If you have male and female turtles, prepare for breeding behavior. Provide nesting sites for females and watch for signs of egg laying. Consult a veterinarian for advice on controlling breeding behavior.

Education provision: sharing knowledge for mutual benefit

Consider participating in educational outreach as a responsible joint snapping turtle owner to share your knowledge and experiences with other hobbyists and the larger community. Raising awareness of proper care, health considerations, and conservation efforts can all help improve the collective well-being of these one-of-a-kind reptiles.

Participate in online forums, social media groups or communities of interest to share knowledge and learn from others. Getting involved in the community can bring valuable insights and help.

Local Workshops and Events: Check out local reptile care and conservation workshops and events. Sharing your information in person can improve other people's understanding and empathy.

Conclusion: Improving the health of common snapping turtles

Understanding the health care needs of common snapping turtles is a complex task that requires dedication, knowledge, and a strong commitment to their well-being. You can provide these ancient reptiles with a satisfying and healthy life in captivity by taking preventive measures, regular health checks and quick response to signs of stress. Your commitment to caring for these unique creatures contributes to broader conservation and conservation initiatives, ensuring the existence of common snapping turtles for generations to come.

Chapter 5

Common turtle handling dos and don'ts

Management of common snapping turtles (Chelydra serpentina) requires a fine balance between respect for their natural behavior and the use of safe and appropriate handling procedures. In this detailed book, we explore the art of managing these ancient reptiles, detailing the dos and don'ts that ensure the safety of the tortoise and the handler.

Understanding turtle behavior: The building block for safe handling

Before getting into the mechanics of handling, it is important to understand the natural characteristics of the common snapping turtle. These reptiles are known for their defensive behavior and ferocious bite that they use to defend themselves. Acknowledging and listening to their feelings is critical to successful and safe treatment.

Defensive Behaviors: Snapping turtles exhibit defensive behaviors when confronted. This includes whistling, lunging and lifting as the name suggests. These are the main defense measures against potential predators.

Territorial Instinct: Common snapping turtles are territorial creatures, especially when it comes to spawning and nesting sites. Management should be provided taking into account their need for personal space.

Aquatic Adaptation: These turtles have excellent aquatic adaptation and are strong swimmers. Consider their comfort in the water and their ability to move quickly through their aquatic environment when handling them.

Common snap turtle do's and don'ts Management: Develop trust and confidence

Successful handling of the Common Snapping Turtle requires patience, respect and attention to best practices. The following characteristics lay the foundation for a safe and positive interaction:

Study the turtle's behavior: Before attempting to capture a snapping turtle, study its behavior. Note his activity level, body language, and any signs of stress. A calm and comfortable turtle can be more patient.

Approach the Common Snapping Turtle Slowly and Gently: When approaching a Common Snapping Turtle, proceed slowly and gently. Quick movements or loud noises can startle them, causing them to react defensively. Instead of approaching from the front, approach from the side.

Use a hand guard: If necessary, use a hand guard, such as a sturdy cardboard or plastic, to guide the turtle in the desired direction. This strategy allows for minimal direct contact while still giving the turtle a sense of security.

When lifting a common snapping turtle, be sure to support the body properly. Hold the turtle by the side of the shell with both hands, avoiding touching the head and tail. Hold firmly so that the turtle does not resist or slide.

Maintain smooth and predictable movements: Maintain smooth and predictable movements while handling. Avoid sudden snaps or quick movements that could startle the turtle. Both the monitor and the turtle benefit from predictability.

Keep in a controlled environment: Keep the turtle in a controlled environment whenever possible, such as a secure outdoor enclosure or separate enclosure. This reduces the chance of the turtle escaping or encountering potential threats.

Provide happy reinforcement: Associate handling with happy experiences by immediately following a handling session with a reward such as a favorite food item. This will build a friendly association, making future relationships more interesting.

Use a blanket for large turtles: A blanket or towel can provide protection for large common garden turtles that are difficult to handle directly. Gently cover the turtle with the blanket, allowing you to pick it up and transfer it without direct contact.

Avoiding concerns and stress when handling common turtles

Avoiding certain actions and behaviors is equally important to ensure the safety of both the handler and the turtle. Avoiding the following when you are pregnant can help reduce stress and potential risks:

Do not grab the tail: Never grab or pull the tail of a turtle that is crawling. Their tails are very sensitive, and rough handling can cause injury. In addition, holding the tail can cause defensive measures, which increases the chance of biting.

Do not catch during nesting season: During breeding season, common loggerhead turtles are known for their tendency to snap. Avoid handling turtles during this time as they can become angry and stressed.

Avoid over-handling: handle only in emergency situations. Repeated or prolonged handling creates stress which can be detrimental to the turtle's health. Allow sufficient time for the turtle to rest and recover between handling sessions.

Do not startle or corner the turtle: Startling or cornering a snapping turtle may trigger an immune response. Instead, approach carefully and avoid backing the turtle into a corner, giving it room to escape.

Avoid direct eye contact: Common snapping turtles may view direct eye contact as a threat. Avoid prolonged eye contact while studying their behavior to reduce anxiety.

Do not handle wild turtles unnecessarily: Wild turtles should be left alone whenever possible. Handling them can disrupt their natural behavior and cause stress.

Contact local wildlife authorities or rehabilitation centers if assistance is needed.

Avoid loud noises: Loud noises can scare snapping turtles, causing them to react defensively. Keep the atmosphere calm and quiet during the treatment to reduce stress.

Handle sick or injured turtles with care: Sick or injured turtles may be weaker and more prone to stress. If you encounter a snapping turtle that appears to be in distress, seek help from a veterinarian or local wildlife officials.

Dealing with problems and troubleshooting: real-world solutions

Dealing with common nuisance turtles can be difficult, even with the best intentions and determination not to do it. Understanding common problems and putting effective solutions in place will help you deal with best practices.

If a sleeping turtle refuses to be handled, look for potential stressors in the environment. Make sure the living space is well designed, with plenty of hiding places and places to hang out. Positive reinforcement should be used gradually to acclimate the turtle to handling.

Biting Tendencies: Biting turtles have a fearsome bite. If a turtle bites while being treated, consider using a portable block or towel to provide a secure barrier. If the biting behavior continues, consult a reptile veterinarian.

Excessive struggling: If the turtle struggles too much during handling, gently return it to its habitat and allow it to rest. Excessive fighting can indicate stress, and pushing the turtle to treat can be harmful.

Aggression Control: Some snapping turtles can become hostile when handled. If hostility is a recurring problem, talk to a behavioral therapist to identify possible causes and make appropriate changes.

Fear responses such as scowling or lunging are common defensive behaviors. Start with short sessions and gradually increase the duration as the turtle gets used to it.

Sharing the responsibility of educational access

Consider participating in educational outreach as a responsible owner or common park turtle enthusiast to share knowledge of acceptable handling techniques. Educating others creates a shared understanding of the needs and experiences of these different reptiles.

Online Forums and Communities: Share your knowledge and experience in online forums, social media groups and communities of interest. Provide guidance on responsible handling procedures and encourage others to consider snapping turtles safely.

Environmental Workshops and Events: Attend workshops or events related to environmental reptiles. Emphasize the importance of demonstrating proper handling skills and respecting the turtles' natural tendencies.

Conclusion: Treat with respect and safety

Caring for common snapping turtles is a delicate ballet that involves respecting their natural habits, their emotions, and carefully studying safe and responsible procedures. By following do's and don'ts, building trust with positive reinforcement, and solving obstacles with practical solutions, you can create an enjoyable and holistic relationship with these ancient reptiles. Responsible handling not only ensures the safety of the keeper and the turtle, but also increases the conservation and appreciation of these exciting species in the wild and in captivity.

Chapter 6

Breeding Common Snapping Turtles: A Pet Owner's Guide

Breeding common snapping turtles (Chelydra serpentina) is a fascinating but complex task that involves careful preparation, knowledge of their reproductive behavior, and dedication to providing suitable conditions for both adult turtles and their offspring. In this comprehensive guide, we'll cover the reproductive organs of common snapping turtles in captivity, from identifying turtle sex to creating the right nesting environment and caring for hatchlings.

Understanding the basics of snapping turtle reproduction

Before starting the breeding process, it is important to understand the basic breeding characteristics of common garden turtles. These turtles exhibit sexual dimorphism, which means that males and females have different morphological characteristics.

Male and female common snapping turtles are easily distinguished by sexual dimorphism. Males have a longer tail, larger claws on their front feet, and a slightly concave plastron (bottom of the shell). Females have flat or slightly curved short tails, small claws and plastrons.

Reproductive Maturity: Common loggerhead turtles reach reproductive maturity at 8 to 10 years of age. However, the age at which they can successfully breed can vary, depending on food, habitat conditions, and individual developmental stages.

Breeding Season: The breeding season for common snapping turtles is normally between spring, April and

June. During this period, turtles become more active, and their mating skills become stronger.

Creating a suitable breeding environment

It is very important to create a suitable breeding environment to encourage natural mating behaviors and effective reproduction. Keep the following criteria in mind when creating a spawning environment:

Aquatic habitat: Make sure the breeding site has a large, well-drained aquatic environment. Snapping turtles prefer clear water, suitable water depth, and a variety of hiding places and caches.

Basking and nesting areas: Include basking areas where turtles can regulate their temperature, such as flat rocks or platforms. Also, provide sufficient nests with sufficient, loose soil or sand for females to lay their eggs.

Temperature Control: Keep the living space at a constant and acceptable temperature. Common snapping turtles are ectothermic, meaning they regulate their body temperature through external heat sources. Ample heating areas with access to sunlight or incandescent lamps are required.

Water Quality: Use a reliable filtration system to ensure good water quality. Regular water changes and water meter monitoring contribute to a healthy breeding environment.

Identification of signs of reproductive behavior
It is important to observe and recognize the signs of breeding behavior in Common Snapping Turtles to determine when they are ready to breed. Common symptoms include:

Increased Activity: During breeding season, both male and female turtles show increased activity. They may spend a lot of time exploring the habitat and engaging in mating behaviors.

Mating Behaviors: Male snapping turtles may exhibit courtship behaviors, including swimming around the female, extending their necks, and gently biting or touching the female's limbs or tail. Reproduction usually occurs in water, the male covers the female.

Nesting Behaviors: Female snapping turtles exhibit nesting behaviors as they prepare to lay eggs. This includes digging a nest in a suitable place, such as sandy soil, and preparing it for laying eggs.

Egg planting and planting

Once mating occurs, females look for suitable nesting sites to lay their eggs. Understanding the process of egg laying and providing the right conditions for successful hatching is essential:

Choosing nesting sites: Female snapping turtles can travel great distances to find suitable nesting sites. Providing a variety of nesting sites within the habitat increases the likelihood of successful egg laying.

Egg Laying: Female turtles can lay anywhere from 20 to 40 eggs depending on the size and age of the turtle. The eggs are round, about the size of ping-pong balls, and have a leathery texture.

Nest preparation: Females carefully dig a nest hole using their hind legs. The depth of the nest can vary, but it is usually several centimeters. After laying the eggs, the female covers the nest with soil, leaves and other debris to hide it.

Incubation Time: The time it takes for turtle eggs to hatch depends on the temperature. Warmer temperatures generally result in shorter incubation times. The eggs are typically left in the nest to develop naturally.

Snapping turtle eggs and hatchlings care

For the success of the breeding project, the time of vaccination and the proper care of the chicks are crucial. Consider the following steps.

Marking and tracking nests: Carefully mark the location of the nests to avoid accidental disturbance. Regularly monitor the nests to ensure they remain undisturbed by predators or environmental conditions.

Protection from predators: Implement measures to protect nests from predators. This may include installing wire mesh or protective barriers around the nest site to protect against predators such as raccoons.

Natural hatching: Allowing the turtle to hatch eggs in a natural nest is often the preferred method. Using eggs or trying to move them can be dangerous and can adversely affect the rods.

Hatching Care: After the eggs hatch, the hatchlings emerge and rise to the surface. Avoid interfering with this process, as hatchlings have special egg teeth that help break the shell.

Safe release: After hatching, it is important to provide a safe and suitable release area for the hatchlings. Make sure the release area has suitable water and land components, allowing the young turtles to mate and explore.

Common problems and troubleshooting

Breeding common park turtles can present challenges, and being prepared to deal with them is essential to a successful breeding project.

Infertile eggs: Some eggs may be infertile, leading to unsuccessful hatching. This may be a natural phenomenon, but if a large number of eggs are infertile, it may indicate issues related to marriage or reproductive health.

Predation Threat: Nests are vulnerable to various animal predators. Apply preventive measures to reduce the risk of premature birth and increase the chance of successful hatching.

Fluctuations in temperature: Fluctuations in temperature can affect the development of eggs and hatchlings. Ensure that the breeding environment provides a stable and suitable temperature.

Hatchling Health: Keep a close eye on the health of your hatchlings. Make sure they have access to proper nutrition, a safe evacuation environment, and protection from potential threats.

Legal considerations and ethical breeding practices
Before starting a breeding project, it is important to be aware of the legal issues and ethical breeding practices associated with common snapping turtles.

Legal Regulations: Check local and national regulations regarding the breeding and ownership of turtles. Some states may have specific requirements or restrictions related to the conservation and breeding of turtle species.

Conservation Ethics: Prioritize conservation ethics when participating in breeding programs. Avoid removing turtles from the wild for breeding purposes, as this can negatively impact wild populations.

Responsible Ownership: If you are purchasing turtles for breeding, make sure they come from reputable sources that prioritize ethical and responsible breeding practices. Avoid supporting illegal trade or harvesting turtles.

Access to Education: Sharing reproductive knowledge responsibly

As a responsible breeder and owner of common snapping turtles, consider participating in an educational service to share your knowledge and experiences. Raising awareness of responsible breeding practices will contribute to the common understanding of these unique reptiles.

Public Education: Share information about reducing turtle reproduction through public talks, educational programs, or online forums. Demonstrate the importance of caring, ethical breeding and responsible pet ownership.

Collaboration with Institutions: Partner with educational institutions, zoos, or conservation organizations to share breeding knowledge and contribute to research or conservation efforts.

Summary: An interesting journey of breeding common snapping turtles

Breeding common snapping turtles is a rewarding but challenging endeavor that requires dedication, knowledge, and a commitment to ethical and responsible practices. By understanding the breeding behaviors of snapping turtles, you can contribute to the conservation and appreciation of these unique reptiles by creating good breeding habitats and solving challenges with practical solutions. As stewards of these ancient creatures, your efforts and responsible ownership play a vital role in ensuring the continued existence of common park turtles for generations to come.

Chapter 7

Common loggerhead turtle species spotlight: unique characteristics

The common snapping turtle (Chelydra serpentina) is a unique and fascinating species that lives in the freshwater habitats of North America. This species occupies a special place in the world of reptiles due to its strong appearance, strong jaws and unique behavior. In this comprehensive focus, we examine the diversity of the common snapping turtle, focusing on its anatomy, characteristics, habitat preferences, and importance in both natural and captive ecosystems.

1. Specific anatomy and physical characteristics
Common snapping turtles have a unique set of physical characteristics that distinguish them from other turtle species. Understanding these traits provides insight into

their evolutionary adaptations and unique characteristics:

Shell Structure: The carapace or upper shell of common snapping turtles is particularly convex along the posterior margin, giving it a rugged appearance. The carapace can vary in color from dark brown to olive or brown and generally ranges from 8 to 18 inches in length.

Plastron Shape: Unlike many turtle species, the common snapping turtle's plastron (lower shell) is small and cross-shaped, leaving most of the body exposed. Plaster is a hinge that allows the turtle to retract its body for protection.

Powerful jaws: One of the most recognizable features of common snapping turtles is their powerful jaws. These turtles have a strong, hooked beak and strong bite force.

The strength of their jaws is adapted to capture prey including fish, amphibians and vertebrates.

Serpentine Neck: Common snapping turtles have a long, flexible neck that resembles a snake, which allows them to strike quickly to capture predators or defend themselves. The neck is often longer in men than in women.

Tail and Limbs: The tails of common snapping turtles are relatively long and can mimic the shape of their necks. Males typically have longer tails than females. In addition, its feet have webbed feet with sharp claws suitable for swimming and digging.

2. Residential preferences and region

Common park turtles are very adaptable and can be found in a wide range of different aquatic habitats. Understanding their habitat preferences and distribution is critical to appreciating their role in various ecosystems.

Freshwater habitats: Common snapping turtles live primarily in freshwater habitats, including ponds, lakes, rivers, streams, marshes, and swamps. They are versatile and can grow in both slow moving and stagnant waters.

Basking Stations: While they spend a significant amount of time in the water, common snapping turtles also rely on basking stations. Perches such as rocks or logs provide opportunities to absorb heat from the sun.

Burrowing Behavior: These turtles exhibit burrowing behavior, create nests for laying eggs and seek shelter during inclement weather or when threatened. Their burrows are located in the soft soil near the water's edge.

Geographic Range: The common snapping turtle has a wide geographic range, covering much of North America. Its distribution extends from southern Canada and the Great Lakes area to the Gulf of Mexico and includes both the Atlantic and Pacific coastal areas.

3. Character traits and adaptations

Behavioral traits of common snapping turtles contribute to their success in a variety of habitats. Understanding their behavior provides insight into their ecology and survival strategies:

Powerful defense: Common snapping turtles are known for their defensive behavior. When threatened, they display violent actions including hissing, lunging and, of course, lifting. Their powerful bite serves as a powerful defense against potential predators.

Nocturnal Activity: Although they can be active during the day, common park turtles are primarily nocturnal predators. They are more active at night, hunting for food and participating in reproductive behavior.

Cautious Behavior: Despite their impressive appearance, common park turtles are often cautious in their interactions. If they recognize potential threats, they can retreat into their shell or water. This careful behavior helps them to avoid unnecessary conflicts.

Territorial nature: Common park turtles exhibit territorial behavior, especially when coming to warmer areas or nesting sites. They can be a protection for these places and can show the attack to the invaders.

4. Reproductive characteristics and life cycle
The reproductive behaviors of common snapping turtles are complex and play an important role in their life

cycle. Understanding their breeding habits provides insight into the dynamics of their populations and the challenges they face:

Mating Ceremonies: During the breeding season, which typically occurs in the spring, male common snapping turtles perform elaborate mating ceremonies. This may involve swimming around females, extending their necks, and gently biting or nibbling.

Nesting Behavior: Females seek suitable nesting sites, often traveling long distances to find good sites. They dig nests in sandy or well-drained soil, lay a clutch of eggs and cover it carefully to protect the nest.

Incubation Time: The incubation time for common snapping turtle eggs depends on temperature. Warmer temperatures generally result in shorter incubation times. The eggs are left in the nest to hatch naturally.

Incubator Pop up Pop up Pop-up is about pop-up. Then they leave the nest and go to the water, where they begin their journey to adulthood.

5. Ecological importance and conservation considerations

Common park turtles play an important role in the ecosystems in which they live, balancing the aquatic food web and participating in nutrient cycling.

Predator-Prey Dynamics: As carnivores, common snapping turtles help control a variety of aquatic organisms, including fish, amphibians, and invertebrates. Their presence affects the dynamics of local ecosystems.

Scavenger Traits: Common snapping turtles are opportunistic scavengers, scavengers and contribute to the decomposition process. This fouling behavior helps maintain the health and cleanliness of aquatic environments.

Indicator species: The presence and health of common snapping turtles can serve as an indicator of the overall well-being of a freshwater ecosystem. Changes in their population or behavior may indicate environmental changes or disturbances.

Conservation Challenges: Despite their adaptability, common snapping turtles face a variety of conservation challenges. Habitat loss, road mortality during nest migration, pollution and illegal harvesting are among the threats to their population.

6. Common snapping turtles in captivity
The fascination of common snapping turtles extends into captivity, where hobbyists and reptile keepers may choose to keep them as pets. Proper care and

understanding of their needs are critical to their well-being.

Adequate enclosures: Captive common snapping turtles require spacious enclosures. Large reservoirs or outdoor ponds can serve as suitable habitats, with adequate hiding places and suitable surfaces.

Nutrition: Providing a nutritious diet is essential to the health of captive common snapping turtles. Their diet should include a variety of protein sources such as fish, insects and occasionally commercial turtle pellets.

Temperature Regulation: Maintaining an appropriate temperature gradient within the enclosure is critical. Thermal access and UVB lamps support thermo regulation, the aquatic environment should have a stable temperature.

Proper Handling: Handling common snapping turtles in captivity requires care and respect for their natural behavior. The use of a hand held block or gloves may be helpful, and contacts should be kept to a minimum to minimize stress.

Summary: A glimpse into the world of common snapping turtles

The common snapping turtle, with its unique characteristics and striking features, is an iconic species in the freshwater ecosystems of North America. From its scaly shell to its powerful jaws, every aspect and characteristic of its body contributes to its success as a versatile and robust reptile.

Understanding the ecological role of the common snapping turtle, its unique adaptations, and the challenges it faces in natural habitats and in captivity creates a deeper appreciation for this species. Whether seen in the wild or kept in captivity by enthusiasts, the common snapping turtle remains a fascinating and important part of the natural world.

As we continue to explore and study these fascinating reptiles, it becomes increasingly important to prioritize conservation efforts, ethical captives, and public education. By doing so, we contribute to protecting and appreciating the common snapping turtle and its vital place in the complex web of life.

Chapter 8

Aquatic Adventures: Setting up an enrichment tank for your turtle

Creating a rich and suitable aquatic environment for your turtle is critical to its safety and overall quality of life. In this comprehensive guide, we'll examine the key elements and considerations in setting up a tank that promotes the physical and mental health of your aquatic turtle.

1. Choosing the right size and type of tank

The foundation of a thriving aquarium environment for your turtle is choosing the right size tank. Tank size should be determined by your turtle's breed and size as well as growth potential. A large tank allows for additional swimming space and the installation of various facilities.

Sizing Tips: As a general rule, the tank should hold at least 10 gallons of water for every inch of turtle shell length. Larger species may require more room. For example, a 60-gallon tank is appropriate for a 6-inch turtle.

Tank Shape: Consider the shape of the tank. Rectangular or long tanks are preferred as they provide a large swimming area. Tall tanks should be avoided as they limit swimming space and are less practical for turtles.

Material: Choose a glass or acrylic tank. These materials are durable, easy to clean and give a great look. Because turtles can be incredibly adept at climbing in and out, make sure the tank is leakproof.

2. Creating a suitable baking environment

Turtles require a specialized basking area to regulate their body temperature, dry off, and gain exposure to UVB light. The cooking area should be comfortable and tailored to your turtle's individual needs.

Baking platform: Build a baking platform in the tank. A floating dock, a floating rock structure, or a platform attached to the tank wall can all be used. Make sure the platform is sturdy enough to support your turtle's weight.

The UVB lamp should be installed above the baking area. UVB light is needed to produce vitamin D3, which helps in calcium metabolism. Use a reptile-specific UVB bulb and replace as directed by the manufacturer.

Maintain the correct baking temperature, usually around 85-90°F (29-32°C). To achieve and maintain the desired temperature, use a heating lamp or a burning lamp. A heat-generating baking lamp is suitable for this.

Substrate: For the nesting area, use a substrate that your turtle can easily climb out of. A comfortable surface can be provided with large, soft river rocks or non-toxic, reptile-safe mulch.

3. Water filter and filter

Maintaining clean, quality water is critical to your aquatic turtle's health. Filtration removes waste, debris and potentially hazardous compounds, resulting in a healthy water environment.

Filter System: Purchase a high quality filter system that is appropriate for the size of your tank. External tin filters are often successful in mechanical and biological filters. Clean and maintain the filter regularly according to the manufacturer's instructions.

Water Heater: Keep the water temperature between 75-80°F (24-27°C). Use an aquarium water heater to keep the water temperature in the recommended range for your turtle.

Water depth: Adjust the water depth according to the size of the turtle. Allow for a deeper pool area and a shallower area for grilling. Using different depths allows the turtles to engage in natural behaviors, and still allow easy access to the nesting area.

Water Testing: Regularly test water parameters including pH, ammonia, nitrite and nitrite levels. Use aquarium test kits to test water quality and make adjustments as needed. Maintaining a stable and clean habitat is critical to your turtle's health.

4. Water decoration and improvement

Enhance your turtle's environment with a variety of water decorations and enhancements. These modifications not only stimulate the brain, but also mimic the elements of their natural environment.

Aquatic Plants: Fill the pool with live or artificial aquatic plants. Live plants can improve water quality by absorbing nitrates and acting as natural hiding places. Make sure any artificial plants you use are made from non-toxic materials.

Hiding Places: Build hiding places out of caves, rocks or driftwood. Turtles like to find a place to rest or hide from the sun. Make sure that the turtle's hiding places are safe and that it does not fall on it.

Turtles often like to sunbathe on floating objects. Floating logs, platforms or cork boards can be added to provide additional collision areas and reinforcement.

Submersible Toys: Include toys and items that can be submerged in water. Turtles can interact with floating objects, impersonate prey, and push objects around. Make sure any objects do not have sharp edges and are safe for your turtle.

5. Consideration for diet and nutrition

A balanced and varied diet is important for your turtle's health and energy. Understanding your turtle's nutritional needs and feeding habits is critical to providing adequate food.

Species-specific diet: Learn about your turtle's food preferences. Some turtles are omnivores, while others are primarily herbivores or carnivores. Offer them a variety of foods that are similar to their natural diet.

High-quality commercial turtle pellets can be used as a primary food source. Look for pellets specially formulated to meet your turtle's nutritional needs. For additional nutrition, supplement fresh and varied foods.

Forage fish, earthworms, crickets and leafy greens are examples of live or fresh foods to include in your diet. These foods are rich in vitamins, minerals and other important nutrients.

Consider using calcium and vitamin supplements to make sure your turtle is getting enough food. Calcium deficiency can be avoided by dusting foods with a calcium supplement.

6. Regular cleaning and maintenance
Regular maintenance is essential to keep the tank clean, prevent the accumulation of hazardous substances, and ensure the overall safety of your turtle.

Do regular partial water changes to remove accumulated dirt and maintain water quality. Change 10-20% of the water every 2-4 weeks, depending on the condition of the tank.

Sub-cleaner: Clean the surface regularly to remove dirt and grime. If you must use a substrate, make sure it is easy to clean and does not pose a risk of contamination.

Filter Maintenance: Clean and maintain the filter system according to the manufacturer's instructions. Clean filter media regularly to prevent blockages and ensure effective water filtration.

Check equipment: Regularly check the operation of equipment such as heaters, filters and lights. Replace damaged or worn parts as soon as possible to avoid cuts around the turtle.

7. Veterinary treatment and health monitoring

Monitor your turtle's health and behavior regularly to catch early signs of illness or stress. A partnership with a veterinarian is critical for routine examinations and treatment of health problems.

Behavior Awareness: Pay attention to your turtle's behavior, appetite, and activity levels. A change in behavior, fatigue, or loss of appetite can all be signs of a health problem.

Annual Checkups: Make an appointment with a veterinarian every year to assess your pet's overall health. Veterinary tests can help identify potential health issues and prompt management if necessary.

When introducing additional turtles or aquatic pals, separate them before bringing them into the main tank. This helps prevent infectious diseases or parasites.

Parasite Prevention: Consult your veterinarian about parasite prevention. Regular stool tests can help identify and treat internal parasites. Follow your veterinarian's parasite prevention and treatment instructions.

Summary: How to care for a healthy and happy turtle

Creating a thriving aquatic environment for your turtle requires careful planning, adequate equipment, and a dedication to their safety. Consider the individual needs of your turtle species, provide adequate bathing and swimming areas, provide a varied and nutritious diet, and maintain a clean and stimulating environment for the overall health and happiness of your aquatic companion.

Remember that each turtle is unique, and observing their behavior and preferences can help you tailor their habitat to their needs. By raising a healthy and content water turtle, you can go on an exciting adventure, building friendships that will last for years to come, with the right care and attention.

Chapter 9

Myths about common snapping turtles: separating fact from fiction

The curious common snapping turtle (Chelydra serpentina) has long been shrouded in myth and misconception. Understanding and appreciating these ancient reptiles requires separating fact from fiction, from their fearsome reputation to their hostility. We dispel common turtle beliefs and provide accurate information about their behavior, biology, and interactions with humans in this in-depth investigation.

1. Misconception: Tortoises are hostile to humans.
Fact:
Although common snapping turtles are known for their defensive habits, they have no innate aversion to humans. Their hateful reputation is often due to their

unique immune system. Snapping turtles may hiss when threatened, open their lips wide, or snap defensively. This is a natural response to perceived dangers and does not directly refer to cruelty to humans.

To avoid potentially aggressive behavior, handle turtles carefully and respect their personal space. Only experienced workers should handle the turtle, and it is very important to consider the safety of the turtle and avoid stress during any interaction.

2. Myth: Snapping turtles can attack by extending their necks.
Fact:
Contrary to popular belief, snapping a turtle's neck has limits and cannot be extended far enough to attack. While snapping turtles have long and flexible necks, their ability to extend them is limited, especially when

compared to the exaggerated descriptions of certain legends.

Snapping turtles primarily use their necks to strike or repel predators. However, their neck extension range is not as wide as it is often seen. Approach turtles with caution and avoid assuming that their necks can reach extreme lengths.

3. Myth: In the water, snapping turtles are always aggressive.
Fact:
Snapping turtles are well adapted to aquatic environments, and although they may show defensive reactions, they are not always aggressive, especially in natural environments. Snapping turtles are more cautious in the water and may prefer to retreat rather than face potential danger.

Snapping turtles do not want human interaction, so meeting them in the wild should be viewed from a respectful distance. Arousing or disturbing turtles in their natural habitat can cause defensive behavior.

4. Myth: Snapping turtles are clumsy and slow on land.
Fact:
While snapping turtles are mostly aquatic, they are incredibly agile and capable of traveling on land, especially when nesting or migrating. Their slow mobility on the ground is frequently interpreted as friction. Snapping turtles may seem slow and brooding, but when on the move, especially at certain times of the year, they can cover large territories.

Female snapping turtles can travel long distances during breeding season to find good nesting sites. During these excursions it is very important to give space and avoid unwanted distractions.

5. Myth: Snapping turtles are omnivorous predators.

Fact:

Snapping turtles are opportunistic hunters, even if they don't point randomly. They have specific dietary needs and appreciate animals such as fish, amphibians, vertebrates and occasionally small mammals. Contrary to popular belief, turtles are selective in their feeding habits, focusing on prey that matches their natural diet.

Understanding turtles' natural behaviors and feeding preferences is critical to coexistence. It is also important to avoid feeding children inappropriate foods, which can lead to malnutrition and health problems.

6. Misconception: Touching turtles is not painful

Fact:

The idea that snapping turtles is painless is a misconception. Turtles, like all creatures, can feel pain and stress. Although they have strong defenses and can withstand harsh environmental conditions, it is not correct to believe that they are painless.

Capturing or engaging in snapping of turtles in a manner that causes unnecessary distress or injury is not only unethical but also harmful to their health. Responsible and courteous behavior should be prioritized to ensure the safety of snapping turtles.

7. Myth: Snapping turtles can easily bite off fingers or toes.

Fact:

While snapping turtles have strong jaws and incisors, the idea that they can easily bite off fingers or toes is an overstatement. Snapping turtles do not have the ability to tear apart important body parts in one bite. Their bite is to capture and eat their prey, not to disperse it.

While caution is important when handling snapping turtles, the risk of losing fingers or legs in a single bite is minimal if proper handling practices are used. Always emphasize safety and avoid exposing any part of your body to the fearsome jaws of a snapping turtle.

8. According to popular belief, snapping turtles are an invasive species in many areas.

Fact:

Common snapping turtles are native to North America and have a wide distribution. In their natural habitat, they are not considered invasive species. But in some areas, they may be considered invaders when they enter outside their homeland.

It is very important to distinguish between natives and invaders. Loggerhead turtles serve critical roles in local ecosystems in their natural habitats, contributing to predator-prey interactions and trophic cycling.

Environmental conservation efforts should be prioritized to protect indigenous peoples and maintain the balance of natural ecosystems.

9. Myth: Petting turtles make good pets.

Fact:

Keeping turtles as pets requires careful planning and following strict care guidelines. The idea that snapping turtles is easy because a pet leads to impulse purchases can cause problems for both the owners and the turtles.

Snapping turtles have specific ecological, nutritional and environmental requirements that must be met in order to thrive in captivity. Caring for turtles as pets involves providing a large and rich home, supporting their appetites, and providing them with adequate bathing and swimming areas. Prospective turtle owners should do extensive research on the breed and be prepared to provide long-term care.

10. Misconception: Snapping turtles threatens fish populations

Fact:

When turtles feed on fish, they are not always a threat to fish species in healthy ecosystems. Snapping turtles help regulate aquatic ecosystems by managing aquatic species such as fish, amphibians, and invertebrates.

Snapping turtles contribute to the overall health and diversity of aquatic populations in a well-balanced ecosystem. Fish population issues are more likely to occur in situations where ecosystems have already been damaged by factors such as habitat loss, pollution or overfishing.

Conclusion: Snapping turtles increase awareness and respect

In order to gain a better understanding of these amazing reptiles, it is important to dispel popular turtle myths.

Tortoises are an important part of North American ecosystems, and understanding their natural habits and ecological responsibilities can help ensure their survival.

Respecting snapping turtles in their native habitat, avoiding unnecessary disturbance, and approaching with caution when encountered are critical steps to coexistence with these ancient and fascinating species. We can raise awareness of the ecological importance of snapping turtles and contribute to their conservation in the wild by dispelling myths and sharing factual information.

Chapter 10

Educational activities for children: common snapping turtles

Common turtle hatching educational activities for kids can be useful and fun. These activities allow young people to learn about these rare reptiles and develop awareness for biodiversity and conservation. The following educational activities, from crafts to interactive courses, are designed to capture the interest of young learners.

1. Build a turtle model for a turtle anatomy lesson
Purpose:
To understand the anatomy and identification characteristics of common tortoises.

Materials:

- Making the play dough or clay shape
- Wood for crafts
- Shaking eyes
- For pipe cleaners
- Markers of different colors

Process:

- Discuss common snapping turtle characteristics such as their shell, legs, tail, head, and distinctive jaws.
- Give each youngster some play dough or modeling clay and tell them to make a model of a turtle.
- Encourage them to use crafts as limbs, pipe cleaners as tails, and googly eyes as eyes for the head.
- Markers can be used to add details such as the scalloped edge of the shell, skin patterns and a distinct beak.

Discussion points:

- How does the turtle's shell protect its body?
- What role do their nails play in their environment?
- Why do you believe snapping turtles have such a unique beak shape?

2. Create an angry turtle environment with a diorama of turtle habitats.

Purpose:

To explore the natural habitat of common turtles and understand healthy ecosystem components.

Materials:

- A shoe box or similar sized box
- Paper for construction

- Dough or modeling clay
- Turtles at least
- Small plants, stones and sticks
- Markers or crayons

Process:

- Discuss common turtle habitats, emphasizing the importance of both aquatic and terrestrial ecosystems.
- Give each youngster a shoebox and teach them how to make a turtle habitat diagram.
- Use construction paper to create sea and land parts in the diorama.
- Add stones, wood and small plants to reflect the natural elements in their habitat.
- To represent snapping turtles, use toy turtles in the diorama.

Discussion points:

- Why do snapping turtles need both water and land in their habitat?
- What habitat features contribute to turtle predation?
- What role do rocks and plants play in their environment?

3. Life cycle storyboard: Draw the life of an angry turtle

Purpose:
To understand the life cycle and developmental stages of common turtles.

Materials:

- Poster board or large sheets of paper
- Colored pencils, markers and pencils

- Pictures of snapping turtles in magazines or paper
- Tape or glue

Process:

- Discuss the typical life cycle of a snapping turtle, including egg laying, hatching, and development.
- Give each child a poster board and ask them to make a story board showing the different life stages of turtles.
- Prepare pictures from magazines or printed materials to represent each step on the storyboard.
- Encourage children to add captions or labels at each stage of the life cycle to explain.

Discussion points:

- How do snapping turtles breed and where do their eggs hatch?
- What happens during the hatching process, and how do young turtles grow?
- How long does it take for a turtle to mature?

4. Trivia Challenge: Snap Turtle Facts Game

Purpose:

A fun and interactive casual game to reinforce knowledge about snapping turtles.

Materials:

- Index cards or paper squares
- Indicators
- container or bowl

Process:

- Prepare turtle-related questions on index cards with varying levels of difficulty.
- Form a circle with the children.
- Explain the rules: each child draws a card from the plate, reads the question aloud and responds to it.
- Record correct answers and distribute points accordingly.
- To improve learning, encourage debate about each question.

Examples of questions:

- What is the main food source for common snapping turtles?
- How do snapping turtles fight back when threatened?
- Where do female turtles normally lay their eggs?

Discussion points:

- Discuss the correct answers and provide additional information on each issue.
- Encourage children to share any fun facts they learned about shelling a turtle.
- Promote the value of understanding and respecting wildlife.

5. Turtle Tales: Storytelling

Purpose:
Through narrative, we hope to inspire creativity by reinforcing information about common snapping turtles.

Materials:

- Notebooks or blank papers
- Colored pencils, pens or markers

Process:

- Discuss important facts about common turtles, including their activities and habitats.
- Ask young people to think of a fictional story in which the main character is a common snapping turtle.
- Encourage children to incorporate what they have learned into their stories while adding creative touches.
- Allow time for young people to write and illustrate turtle stories.
- Allow them to share their stories with the group, creating a creative and educational storytelling session.

Discussion points:

- Have each child share the main facts they included in their story.

- Discuss the creative elements and innovative features of each narrative.
- Emphasize the importance of combining empirical knowledge with creative history.

6. Protection poster: Protecting snapping turtles

Purpose:

To raise awareness about the importance of conservation and the role children can play in saving common turtles.

Materials:

- Poster board or large sheets of paper
- Colored pencils, markers and pencils
- Conservation-related magazines or printed images
- Tape or glue

Process:

- Discuss the conservation status and challenges of common tortoises such as habitat loss and road mortality.
- Explain the importance of conservation efforts and how individuals, including children, can contribute.
- Provide each child with a poster board and teach them to create a conservation poster highlighting ways to protect snapping turtles.
- Encourage the use of images, slogans and creative elements to communicate the conservation message.

Discussion points:

- Discuss the risks of poaching turtles and the impact of habitat loss and road mortality.

- Explore ways kids can contribute to snapping up turtle conservation efforts.
- Emphasize the importance of responsible behavior in natural ecosystems.

7. Outdoor exploration: Turtle tracking adventure

Purpose:
Connect children to nature by exploring outdoor settings and observing vibrant turtle ecosystems.

Materials:

- Binoculars (optional)
- Clipboards
- Pens or pens
- Nature notebooks or observation sheets

Process:

- Take the kids on an outdoor adventure to a local pond, lake or wetland.
- Provide each youth with a clipboard, pencil, and nature notebook or observation paper.
- Teach them to look around for signs of turtles or their habitat shrinking.
- Encourage children to list any plants, rocks, or snags that might attract snapping turtles.
- Talk about your observations and experiences from the outdoor adventure.

Discussion points:

- Discuss the importance of respectfully avoiding wild animals.
- Encourage children to share their discoveries and signs of snapping turtles they find.
- Explore ecosystem interdependence and the role that turtle poaching plays in maintaining balance.

Finally, encourage curiosity and protection

Children have a unique opportunity to learn about wildlife, conservation, and ecosystem interconnectedness through educational activities focused on common turtles. These activities aim to build curiosity, respect and responsibility through a mix of purposeful projects, interactive games and outdoor adventures.

Kids can engage with the fascinating world of snapping turtles while reinforcing critical educational concepts through creative projects like making turtle models, creating habitat dioramas, and writing turtle stories. Additionally, activities such as trivia quizzes and outdoor exploration encourage active participation and experiential learning.

These educational activities help build environmentally conscious individuals who instill a love of the outdoors and wildlife and appreciate the importance of conservation. Finally, connecting young people with common snapping turtles fosters a sense of stewardship and a commitment to protecting the many ecosystems these amazing reptiles call home.

Questions to be asked

What is a common snapping turtle?

A large freshwater turtle found in North America is the common snapping turtle (Chelydra serpentina).

Q: What is the maximum size of a common snapping turtle?

A: Common small turtles can have a shell length of 18 inches or more.

Q: How long does a common snapping turtle live?

A: They can live for decades in the wild, with some individuals living for 30 years or more.

Q: Where can you find common small turtles?

They are found in freshwater in North America from southern Canada to Florida.

Q: What makes snapping turtles different from other species?

A: Snapping turtles have strong jaws and a long, flexible neck, as well as a distinct, hooked beak.

Q: Do snapping turtles make good pets?

A: Because they require specialist care and large habitats, they are not suitable for inexperienced reptile keepers.

Q: What kind of food do common snapping turtles eat?

A: They eat fish, invertebrates, amphibians, small animals and aquatic plants.

Q: Do turtles eat fruits and vegetables?

A: Although they are mostly carnivores, they may eat some fruits and vegetables.

Q: What defenses do snapping turtles have?

A: They fight their prey by snapping, gnawing and emitting a foul smell.

Q: Do snapping turtles sunbathe?

A: They warm up to regulate their body temperature and absorb sunlight.

Q: Are snapping turtles dangerous?

A: They can be defensive, especially when threatened, but they are not naturally aggressive towards humans.

Q: Can snapping turtles coexist?

A: They are primarily solitary creatures with territorial tendencies.

Q: What is the purpose of a turtle's long tail?

A: The tail is used for swimming and is used as a reproductive organ in females.

Q: How do snapping turtles meet?

A: Mating involves courtship behavior, and females lay eggs in nests dug in sandy or loose soil.

Q: Do snapping turtles make good tanks with other turtles?

A: They can be hostile to other turtles, especially if space is limited.

Q: Can snapping turtles be kept in a small tank?

A: They need a wide enclosure with both water and land environments.

Q: How deep should the water be in a turtle tank?

A: The depth of the water should allow for swimming, and a deep range is necessary for swimming.

Q: Can snapping turtles be raised in outdoor ponds?

A: Yes, but outdoor enclosures must provide a safe and predator-proof environment.

Q: Do snapping turtles sleep?

A: Yes, they can hibernate in the winter by burying themselves at the bottom of bodies of water.

Q: How often should I feed my snapping turtle?

A: Feed juveniles daily and adults every 2-3 days, adjusting for age and activity level.

Q: Can I hand feed my snapping turtle?

A: Not recommended as their feeding response is fast and they may confuse their fingers for food.

Do Turtles Need UVB Lamps?

A: Yes, UVB exposure is required to produce vitamin D3 for normal calcium metabolism.

Q: How do I handle a cracked elk?

A: Be careful, keep your hands away from your head, and consider wearing gloves or protection.

Q: Do snapping turtles come in different varieties?

A: Yes, there are different species, including the common snapping turtle.

Q: Do snapping turtles have the ability to climb?

A: They are not skilled climbers, although they may attempt to climb obstacles in some situations.

Q: What are the most common health problems in snapping turtles?

A: Common health issues include respiratory infections, shell injuries and nutritional deficiencies.

Q: How can I tell if a snapping turtle is male or female?

A: Males have long tails and broad shells, while females have short tails and narrow shells.

Q: Do snapping turtles drop shells?

A: No, shells are not discarded like skins; Instead, they grow together with the turtle.

A: Snapping turtles can live in shallow water.

A: They are mostly freshwater turtles and do not do well in stagnant water.
Q: Are there any hunters who take turtles?

A: Raccoons, birds, and large aquatic predators prey on eggs and young turtles.

Q: Do snapping turtles have the ability to retreat into their shells?

A: Tortoises, unlike many other turtle species, cannot fully retract their shells.

Q: How often should I clean my snapping turtle's tank?

A: Regular partial water changes and substrate cleaning are required; The frequency varies according to the size of the tank.

Q: Can turtles be trained to snap?

A: They are difficult to teach, but they can be taught to associate their owner with food.

Q: Are there legal provisions for keeping snapping turtles as pets?

A: In some states, keeping turtles may be prohibited, so check local laws.

Q: Do snapping turtles have the ability to recognize their owners?

A: They recognize their owners as a source of food, but do not develop a significant relationship with them.

Q: Can snapping turtles communicate?

Answer: They may hiss when threatened but are not vocal.

Q: Can snapping turtles live in cold weather?

A: They can withstand mild temperatures, but in extreme cold, they can be weak.

Q: How do snapping turtles move in water?

Answer: They are strong swimmers that propel themselves with their webbed feet.

Q: Can turtles be kept in the same aquarium as fish?

A: Catching turtles is not recommended as they may consider fish as predators.

Q: Does snapping turtles only eat turtle pellets?

Answer: Although supplements can be part of their diet, they should eat a varied and balanced diet.

Q: Do snapping turtles carry viruses that are dangerous to humans?

A: Although they can carry Salmonella, proper sanitation procedures will limit the risk of transmission.

Q: Can you find snapping turtles in cities?

A: They can live in ponds, lakes and other water bodies in cities and suburbs.
Does a snapping turtle migrate?

A: Some individuals may perform seasonal activities, especially during breeding season.

Q: Can snapping turtles survive their entire lives in captivity?

A: Turtles can be raised in captivity with enough care, but dedication is required.

Q: How can snapping turtles communicate with each other?

A: Much of communication is non-verbal, consisting of body language and defensive behaviors.

A: Are the turtles in danger of snapping?

A: They are not currently endangered, although habitat loss threatens some populations.

Can turtles eat small mice?

A: When feeding small mammals, a balanced diet is important for their well-being.

Q: Does snapping turtles smell good?

A: They have a developed sense of smell, which they use to find food.

Q: How do I give my snapping turtle a boost?

A: Provide different areas with hiding places, hiding places, and things to investigate.

Can snapping turtles live in saltwater?

A: No, they are freshwater turtles that cannot grow in salt water.